CITRUS
A Cookbook

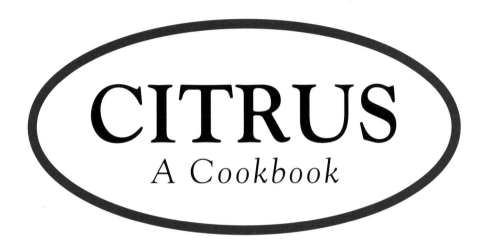

CITRUS
A Cookbook

GREENWICH EDITIONS

A QUANTUM BOOK

This edition published by
Greenwich Editions
10 Blenheim Court
Brewery Road
London N7 9NT

Copyright © 1992 Quarto Publishing plc

This edition printed 1997

ISBN 0-86288-155-2

QUMCTU

This book was produced by
Quantum Books Ltd
6 Blundell Street
London N7 9BH

Printed in Singapore by Star Standard Industries Pte Ltd

For my mother and father, my grandmother Lois Everson, and Ruth
Boston, who taught me to love food

ACKNOWLEDGMENTS

Special thanks to Steve Smith for his inspiration, support and
love, and for transferring my scribbles to disk. For their love
and encouragement, thanks to my family and friends, particularly
Martie LaBare and Michael McEvoy, Fred and Beth Rust, and John
Morse for tasting recipes, and Alan Mittelsdorf for his
friendship and support over the years.

Thanks to Carla Glasser and John Morse (again) at the Nolan-
Lehrer Group for getting me started. Special thanks to Chris Fagg
at Cassell, and Marta Hallett, Ellen Milionis, Jill Hamilton, and
Rose K. Phillips at Running Heads, and Linda Greer and Norma
MacMillan, as well. Finally, thanks to Dexter Samuel for his good-
humored help and to David Blackburn, Jeff Romano, and Tina Mae
Jones of Native Farms for sensational citrus.

We would like to thank the following people for their support
during this project: the Jacobson and Ehlers families, Judy
Devine for her inspiration and love, Dexter Samuel, Jill Bock,
Beth Farb, Michele Cohen, Kim Kelling, Tina Klem, Jon Roemer,
Robin Van Loben Sels, and Linda Winters.

Thanks also to the many people and companies who loaned us their
beautiful tablewares: Williams-Sonoma, Pottery Barn, Laura Beck
at Platypus, Kaija at Ceramica, Jim Mellgren at Dean and Deluca,
Pauline Kelley and Lorena Sita at Zona, John P Gilvey and Michael
Benzer at Studio Art Glass, Kathy Dahlberg at Bayer Glass
Studios, and Elizabeth MacDonald for her fabulous tilework.

INTRODUCTION
10

ONE
BREAKFAST AND
BRUNCH

Pink Grapefruit Poached
in Sauternes
20

Stewed Kumquats and
Strawberries
22

Orange and Banana Muffins
24

Lemon and Allspice Muffins
24

Tangerine and Pecan Scones
26

Navel Orange Waffles with
Blueberry Sauce
28

Orange Pain Perdu
30

Buckwheat Pancakes
with Clementines
32

TWO
STARTERS, SOUPS AND
SALADS

Grilled Pink Grapefruit and
Pork Skewers
36

Lime Soup
38

Lemon and Parsley Soup
40

Curried Orange Soup
42

Salad of Beetroot, Lamb's
Lettuce and Clementines
44

Ugli Fruit Caribbean Fish Salad

46

Chicken Salad with Tangerines

and Red Onions

48

Pickled Citrus Prawns

50

THREE
MAIN COURSES

Lemon Fettuccine with

Peppered Prawns

54

Fried Scallops with Lime

and Garlic

56

Grilled Shark Steaks with

Citrus Salsa

58

Baked Fish with Ugli Fruit

60

Grapefruit-Marinated Poussins

62

Chicken with Lemon and Olives

64

Soba with Fennel, Chèvre

and Kumquats

66

Spareribs with Lime and Chilli

68

Pork Medallions with Mandarin

Oranges and Cranberries

70

Veal Chops with Blood Oranges

72

Beef with Tangerines

74

FOUR
RELISHES AND
CONDIMENTS

Pickled Lemons
78

Citrus Marmalade
80

Kumquat and Ginger Preserve
82

Lime and Tomato Relish
84

Tangelo and Onion Relish
84

Tangerine and Cranberry
Chutney
86

Orange Curd
86

FIVE
DESSERTS

Maple and Orange Ambrosia
90

Chocolate-Dipped Citrus
Segments
92

Minted Ruby Grapefruit Ice
94

Tangerine Sorbet
96

Lemon Snaps
96

Chocolate and Orange
Marmalade Brownies
98

Orange and Coconut Custard
100

Lime Mousse Tart
100

Frozen Blood Orange Soufflé
104

Lemon and Nutmeg Cheesecake
106

Lemon Steam Cake with
Blood Orange Sauce
108

Mandarin Orange Upside-
Down Cake
110

Hot Spiked Lemonade
118

Lemon and Thyme Tea
118

Sherried Pink Grapefruit Juice
120

Grapefruit and Pineapple
Rum Punch
120

Blood Orange Liqueur
122

SIX
BEVERAGES

Pink Lemonade
114

Iced Orange Coffee
114

Limeade Fizz

116

Citrus Frappé
116

RECIPE LIST
124

RECIPE LIST BY CITRUS
126

INTRODUCTION

To many people, the uses for citrus fruits extend only to drinks and to a few familiar dishes. For centuries, however, cooks in cultures all over the world have found that the ways to use oranges, lemons, grapefruits, limes and more unusual varieties such as ugli fruit and kumquats – are practically limitless.

Growing up in Florida, both a citrus and seafood centre, I learned, early on, the indispensability of citrus when serving all kinds of seafood and that serving iced tea without a slice of lemon is tantamount to blasphemy. As a child, I travelled to Key West for my first real Key lime pie and was surprised to find it was yellow instead of green. I have vivid memories of climbing my family's garden orange tree, which would fill the air with the sweet scent of its blossoms in springtime. My family would trade our oranges for the giant yellow grapefruits our next door neighbours grew, and we always had plenty of lemons on hand for lemonade to make the hot and humid summers more bearable. Like many kids, my first commercial venture – the lemonade stand – owed everything to this versatile crop. Fresh citrus fruits formed the basis of the many fruit salads, refreshing main dishes, and even the home remedies (my mother favoured grapefruit 'punch' for laryngitis) we enjoyed year-round.

As I got older, it didn't surprise me to find that other cultures carried the love of citrus even further. I've enjoyed my encounters with citrus in cuisines from around the world. Caribbean cooks offer such wonderful dishes as seviche and a citrusy Christmas pudding, not to mention their citrus-infused signature planters punch and tropical cocktails. Grecian fare makes ample use of the lemon, in their avgolemono soup and such staples as squid with lemon and garlic. Asian cultures, too, rely on citrus to add diversity. Almost everyone is familiar with beef with orange, but there are also many other interesting citrus dishes to enjoy, such as Thai lime yum dressing (yum is a type of salad). Marvellous and inventive dishes such as these have inspired me to create the citrus recipes for this book.

Archeobotanists believe that a common ancestor of all of today's citrus fruits sprang up in India, or possibly in the Tigris and Euphrates Valley, 80 centuries ago. Crossbreeding refined and differentiated various forms of citrus, and these were further distributed through the known world. As this happened, the types of fruit and their uses became as varied as their locales.

Grown in Mesopotamia for their beauty and scent and in Egypt for their use in embalming, citrus fruits have been used at various times as aphrodisiacs, as cures for fever or colic, or as a means of protection against poisons. Romans used a certain variety of citrus to keep moths from their woollens, while they relied on sour grapes to add tartness to their food (until Persian slaves introduced them to citrus's culinary uses). In the Dark Ages – a time when eating fresh fruit was considered by many to be harmful to the body – lemons were thought to be poisonous.

The discovery of scurvy prevention through citrus was made long before it was understood. In fact, the first citrus in the Western Hemisphere probably arrived with Christopher Columbus. The seeds quickly spread throughout the islands of the Caribbean, making the islands medical way-stations for sailors. By the time Ponce de Leon arrived in Florida in 1513, sailors on Spanish ships were required to carry 100 citrus seeds for planting wherever they landed. By the 19th century, citrus had become a major industry in Florida.

Among modern day inventive uses for citrus that I've encountered are employing lemon juice as invisible ink and the old fisherman's trick of using lemon juice to remove fishy smells after cleaning the catch of the day. Citrus flowers, with their gorgeous scents, are potent ingredients in pot-pourri and other fragrant household blends.

But the most rewarding way to use citrus is, without question, as a culinary ingredient. You'll find recipes here that use navel and blood oranges; white, pink, and ruby red grapefruits; tangerines or mandarin oranges; and lemons, limes, clementines, tangelos and kumquats. Today, many of these fruits are available year round, but others are still seasonal treats to be searched out and enjoyed but briefly. All are abundant in vitamin C.

Oranges in one variety or another are with us all the time. Some varieties, such as blood oranges, are usually found only in the winter and spring; others, like navels, are available most of the year. Valencias, a late variety, are in the shops from the end of spring throughout the summer.

When selecting oranges (as with any citrus fruit) look for smooth firm skins, free of soft spots. Pick a fruit that feels heavy for its size.   The exception to the smooth-skin rule is the unfortunately named 'ugli fruit'. I think of it as the Shar Pei of citrus fruits, with a loose and wrinkled skin that hides what is, perhaps, the best-kept secret on the citrus market. Uglis are sweet, delicious and just about the juiciest fruit around. Originally a hybrid of a grapefruit and a tangerine, the ugli grows in the Caribbean and is available from December to May.

One of the ugli's ancestors, the tangerine, is also called a mandarin orange. Though the terms are interchangeable, smaller tangerines often end up labelled 'mandarins'. Tangerine skin doesn't have the bitter layer of white pith that surrounds other citrus fruits, but rather a few strands of it between the skin and the fruit. This makes the skin of tangerines particularly delightful and useful in cooking, such as with Tangerine and Pecan Scones. Throughout the book, the term zest is used to describe the coloured part of the peel; however, with the tangerine, the entire peel is used. Of all citrus fruits, the tangerine's skin tastes most like the fruit it encloses, and it doesn't need the blanching often required with the others to remove the bitter taste. This delightful fruit is sweet and tasty, as Chinese cooks have known for centuries, and can be used fresh or dried. When selecting tangerines, look for those with deep colour and loose skin for easier peeling.

The ugli fruit's other ancestor, the grapefruit, has had a hard time escaping its typecasting as a breakfast fruit. Available in traditional white and pink, as well as the more recent ruby red variety, it is a far more versatile fruit than many people believe. Grapefruit makes thoroughly thirst-quenching drinks and palate-cleansing ices, as well as being a spectacular salad ingredient. If you insist on relegating grapefruit to the breakfast table, try a little salt on it instead of sugar. To my own taste, I find it sweet enough – particularly the ruby reds with no additions. You can probably find grapefruit in your shops all year, but they'll be freshest and in greatest abundance from January to April. Choose fruit that has a nice lustre and is well-shaped and heavy.

Clementines are a cross between mandarins and bitter Seville oranges, but the taste is anything but bitter. They are sweet and easy to peel and segment, and, best of all, seedless. They're the 'little darlings' of the citrus trade. Unfortunately, clementines are available for only a relatively brief period in the winter. Smaller even than the clementine is the kumquat which some consider the stepchild of citrus, with its reputation as a sour pucker-upper. These days, through cultivation, this stepchild seems to be something of a 'Cinderella'. It can be surprisingly sweet, particularly the skins. Some people even eat them for the purpose of stopping cravings for sweets. Traditionally limited to preserves or candied, the kumquat is coming into its own by lending its piquant flavour to fish and vegetable dishes. Buy kumquats that are firm and shiny, without blemishes or soft spots. If the green leaves are still attached, that's a good clue to the fruit's freshness.

Lemons and limes are available all year, but their prices vary with the season. Avoid shrivelled fruits or those with very soft or very hard skins. Lemons with green-tinged skins and the greenest limes are the tartest.

You'll find many varieties of lemon today, but the small round thin-skinned ones tend to be juiciest. Limes, which require hotter moister climates to thrive, are used in place of lemons in many parts of the tropics. There are two basic types of lime you'll find in the shops: the small, yellowish Indian variety, and the larger, very green Caribbean one.

In general, I prefer to use organic fruit when possible, particularly when using the peel or zest. If you use non-organic fruit, which may have been sprayed with a fungicide or waxed, always wash it thoroughly in warm soapy water and rinse well. Complaints about organic fruits being less attractive just don't hold up anymore. Of course, some fruits do appear preternaturally bright and colourful. And that may well be because they've been dyed. Usually, Florida citrus isn't dyed, and a bit of greenish colour is acceptable.

Here are a few more tips for getting the most out of your ever-versatile citrus selections:

- Most citrus fruits will keep at room temperature for 2 to 5 days. They'll keep in the refrigerator for 2 to 4 weeks.
- Use lemon juice in place of salt (this is especially good for those on a low-salt diet) or in place of vinegar for a cool clean taste.
- Add the juice of a lemon, instead of salt, to the water used to boil pasta.
- To avoid discoloration, add citrus juice (lemon is usually used, but any fruit with citric acid will do) to lightly coloured fruits such as apples, bananas, pears or avocados, when cutting them up for serving or freezing.
- To prevent darkening, add any kind of citrus juice to the cooking water of lightly coloured vegetables such as cauliflower, turnips or potatoes.
- Substitute 1 teaspoon of lemon juice for $1/2$ teaspoon of cream of tartar in meringue recipes.
- Tenderise tough meat by marinating it in any kind of citrus juice or by spooning juice over the meat which has been pierced with a fork.
- Freshen poultry after washing by rubbing it with a cut lemon or orange.
- Add 1 teaspoon of lemon juice to each 450 g (1 lb) of fruit when bottling, if you are unsure of the fruit's acid content.
- Never use iron or aluminium cookware in preparing citrus, as it will react with the metal, resulting in a metallic taste and darkening of the food.

Citrus fruits can be found wherever warm winds blow and the tropical sun shines. For the rest of the world, these fruits are tropical ambassadors with their bright colours and fresh, cool tastes. They have become integral ingredients in the cuisines of the world. Some dishes such as lemon meringue pie, duck a l'orange and lime sorbet are very well known and have appeared in hundreds of cookery books. You may not find these and other old favourites in this book, but you might make a new friend in these sometimes surprising ways for using citrus fruit in cooking.

I think you'll find citrus's clean cool tastes - sometimes tart and sometimes sweet - infinitely adaptable. Citrus fruits can enhance virtually any other flavour without overwhelming it. They can be used with the most delicate tastes such as chicken or cucumber where they support the flavour, or with pronounced tastes such as lamb or onion where they come into their own, delightfully altering the other familiar tastes.

My sources for these recipes come from all over the world, and from my own experience and imagination. In all cases, though, my inspiration comes from the simple to use, delicious to taste citrus fruits.

"Know you the land where the lemon trees bloom?
  In the dark foliage the gold oranges glow;
  a soft wind hovers from   the sky...
  There, there I would go, my beloved, with thee!"

GOETHE
Wilhelm Meisters Lehrjahre

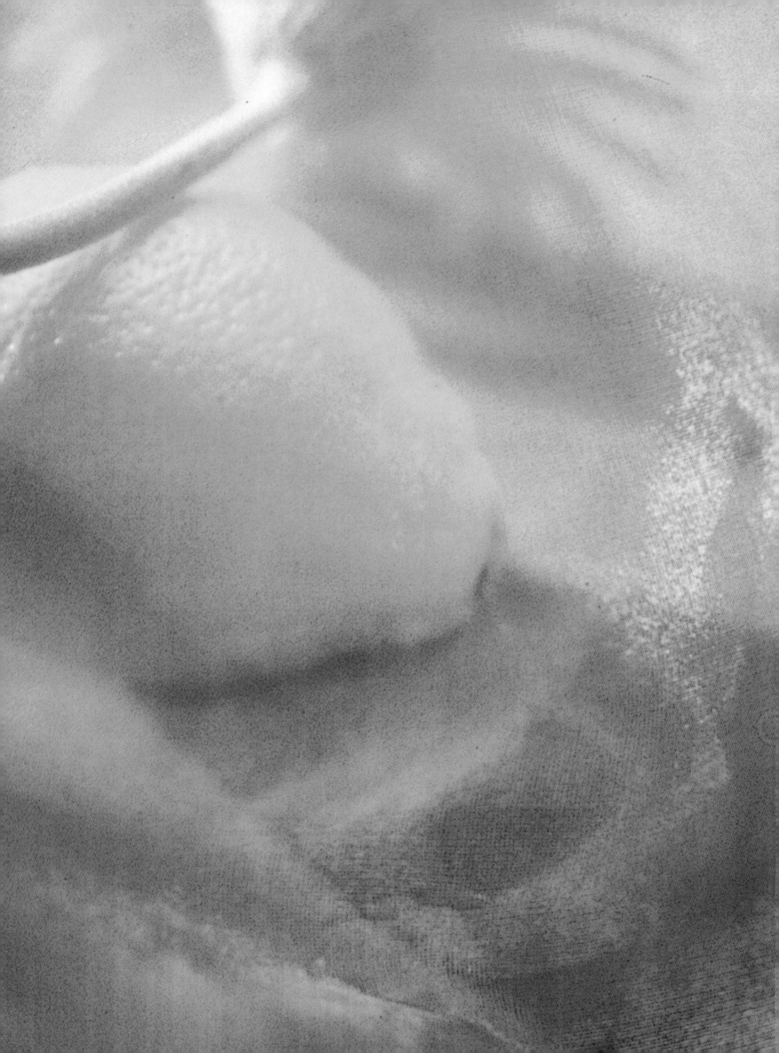

ONE
BREAKFAST
AND BRUNCH

PINK GRAPEFRUIT
POACHED IN SAUTERNES

3 pink grapefruits

500 ml (16 fl oz) Sauternes

3 tablespoons brown sugar

mint leaves to garnish

Serves 4

Preparation time: 15 minutes

Using a small sharp knife, peel the grapefruits over a small bowl, cutting through the peel and pith, just into the flesh, then spiralling down and around from one end to the other, as if peeling an apple. Cut into the membranes, pop out the segments, and remove the pips.

In a large non-reactive saucepan, heat the Sauternes and sugar almost to boiling. Add the grapefruit segments and simmer for 5 minutes. Remove from the heat and allow the grapefruit to cool a bit in the Sauternes. Serve in the sauce while still slightly warm, or chill for serving later. Garnish with the mint leaves.

STEWED KUMQUATS AND STRAWBERRIES

350 g (12 oz) kumquats, washed and stalks removed

150 g (5 oz) caster sugar

5 cloves

350 g (12 oz) strawberries, washed and hulled

2 tablespoons Grand Marnier

Serves 4

Preparation time: 30 minutes, plus several hours or overnight to chill

Slice each kumquat crossways into 3 or 4 rounds. Remove pips. Place in a large non-reactive saucepan with the sugar, cloves and 125 ml (4 fl oz) water. Simmer for 5 minutes.

Quarter the strawberries and add them to the pan. Simmer for 3 more minutes, then remove from heat. Allow to cool to room temperature. Add the Grand Marnier; cover and refrigerate for several hours or overnight. Remove the cloves before serving. Serve as is or over vanilla ice cream.

ORANGE AND BANANA MUFFINS

90 g (3 oz) plain wholemeal flour

60 g (2 oz) rolled oats

75 g (2½ oz) unbleached plain flour

1 tablespoon baking powder

75 g (2½ oz) soft brown sugar

½ teaspoon salt

3 tablespoons wheat germ

1 teaspoon ground cinnamon

125 ml (4 fl oz) milk

125 ml (4 fl oz) freshly squeezed orange juice

1 size-3 egg, lightly beaten

2 bananas, mashed

2 tablespoons grated orange zest

60 g (2 oz) unsalted butter, melted

Preheat oven to 200ºC (400ºF or gas 6).

Grease six deep 7.5 cm (3 in) bun tins. In a large bowl, combine the dry ingredients. In a small bowl, combine all the remaining ingredients, mixing well. Add the liquid ingredients to the dry ones all at once and stir just until moistened. Spoon the mixture into the prepared tins, filling them ³/4 full.

Bake for about 25 minutes, or until they are golden brown and a wooden cocktail stick inserted in the centre of a muffin comes out clean.

Makes six 7.5 cm (3 in) muffins
Preparation time: 45 minutes

LEMON AND ALLSPICE MUFFINS

300 g (10 oz) unbleached plain flour

2 teaspoons baking powder

½ teaspoon salt

100 g (3½ oz) caster sugar

1½ teaspoons ground allspice

1 tablespoon grated lemon zest

1 size-3 egg, lightly beaten

150 ml (¼ pint) milk

5 tablespoons freshly squeezed lemon juice

60 g (2 oz) unsalted butter, melted

Preheat oven to 200ºC (400ºF or gas 6).

Grease twelve 6 cm (2½ in) deep bun tins.

Sift the flour, baking powder, salt, sugar and allspice into a mixing bowl. Add the zest and mix well. In a small bowl, combine the beaten egg with the milk, lemon juice and melted butter.

Add the liquid ingredients to the dry ones all at once, mixing only until well moistened (the mixture should be a little lumpy). Fill bun tins ³/4 full with the mixture. Bake for 20 to 25 minutes, until golden brown.

Makes twelve 6 cm (2½ in) muffins
Preparation time: 45 minutes

TANGERINE AND PECAN SCONES

300 g (10 oz) unbleached plain flour

50 g (1³/₄ oz) caster sugar

1 tablespoon baking powder

¹/₂ teaspoon bicarbonate of soda

¹/₂ teaspoon salt

120 g (4 oz) unsalted butter, frozen and cut into chunks

peel from 2 to 3 tangerines, all white pith strands removed

90 g (3 oz) shelled pecan nuts, coarsely chopped

175 ml (6 fl oz) buttermilk

1 size-3 egg yolk

Makes twelve 7.5 cm (3 in) scones
Preparation time: 40 minutes

Preheat oven to 200ºC (400ºF or gas 6).

In the bowl of a food processor or by hand, combine the flour, sugar, baking powder, bicarbonate of soda, salt and butter. Process until the mixture resembles coarse crumbs. Transfer to a bowl. Dice the tangerine peel into 5 mm (¹/₄ in) pieces.

Add the peel and pecans to the mixture. Add the buttermilk and mix well, until the dough is stiff enough to handle. Turn out on to a well floured board and knead gently until smooth, about I minute. Roll or pat out the dough to 2 cm (³/₄ in) thickness. Cut with a 7.5 cm (3 in) biscuit cutter or drinking glass dipped in flour. Arrange scones on greased baking sheet 2.5 cm (1 in) apart. Re-form dough scraps into a ball, roll out and cut again until all dough is used. Beat egg yolk with 1 teaspoon of water and brush over tops of scones. Bake for 15 to 20 minutes, until golden brown.

NAVEL ORANGE WAFFLES WITH BLUEBERRY SAUCE

250 g (8¹/₂ oz) unbleached plain flour

1 teaspoon baking powder

1 teaspoon bicarbonate of soda

¹/₂ teaspoon salt

2 tablespoons caster sugar

400 ml (14 fl oz) buttermilk

60 g (2 oz) unsalted butter, melted

3 tablespoons Cointreau or ¹/₂ teaspoon orange flower water

2 size-3 eggs

2 tablespoons grated navel orange zest

vegetable oil

navel orange segments

Makes 4 to 6 waffles, depending on iron size

Preparation time: 30 minutes

Preheat waffle iron.

In a large bowl, combine the flour, baking powder, bicarbonate of soda, salt and sugar. Stir to mix. Add the buttermilk, butter, Cointreau, eggs and zest and beat until smooth.

When the waffle iron is ready, brush it lightly with oil and pour batter into the centre until it spreads within 2.5 cm (1 in)of the edge. Cover and bake following the manufacturer's instructions.

When done, carefully lift the cover and loosen the waffle with a fork. Serve immediately, garnished with orange segments and Blueberry Sauce (recipe follows), or keep warm, uncovered, on a baking sheet in a preheated low oven, until all waffles are cooked.

BLUEBERRY SAUCE

300 g (10 oz) blueberries, washed and stalks removed

100 g (3¹/₂ oz) caster sugar

1 tablespoon freshly squeezed lemon juice

In a small saucepan over medium heat, combine the blueberries with sugar, lemon juice, and 4 tablespoons water. Bring to the boil and simmer for 5 minutes. Serve warm.

ORANGE PAIN PERDU

4 tablespoons soured cream or plain yogurt

1 teaspoon ground cinnamon

2 tablespoons honey (preferably orange blossom honey)

2 size-3 eggs

300 ml (½ pint) freshly squeezed orange juice

1 loaf French, challah or Italian bread, 2 to 3 days old (the harder and crustier the better), cut into 2.5cm (1 in) slices

unsalted butter for frying

Serves 4

Preparation time: 15 minutes

In a small bowl, whisk the soured cream or yogurt with the cinnamon and honey. Add the eggs and beat until smooth. Gradually beat in the orange juice and continue to whisk until well combined and frothy. Dip the bread slices into the mixture a few at a time, soaking them thoroughly, but not so much that they fall apart. Melt 45 to 60 g (1½ to 2 oz) of butter in a frying pan over medium-high heat and sauté the slices until golden brown on each side. Do not crowd the slices in the pan. Add more butter as needed.

Serve immediately or keep warm in a low oven until all slices are cooked. Serve with Orange and Maple Syrup (recipe follows) or honey.

ORANGE AND MAPLE SYRUP

250 ml (8 fl oz) maple syrup

4 tablespoons freshly squeezed orange juice

In a small saucepan over low heat, combine the ingredients, stirring often, until heated.

BUCKWHEAT PANCAKES WITH CLEMENTINES

100 g (3$^1/_2$ oz) buckwheat flour

75 g (2$^1/_2$ oz) unbleached plain flour

$^1/_4$ teaspoon salt

400 ml (14 fl oz) milk

45 g (1$^1/_2$ oz) unsalted butter, melted

3 size-3 eggs

4 clementines

75 g (2$^1/_2$ oz) caster sugar

350 g (12 oz) mascarpone cheese, at room temperature

vegetable oil

Serves 4. Makes 12 pancakes

Preparation time: 1$^1/_2$ hours

Sift the flours and salt into the bowl of a food processor or electric mixer. Process or beat for a few seconds to combine. With motor running, add the milk, butter and eggs. Process just until smooth and blended. Set the batter aside for 30 to 45 minutes. Peel the clementines, removing all the white pith and reserving the peel of one of them. Separate and reserve the segments.

Scrape the white pith off the reserved peel, then finely chop it. Place it in a small saucepan with the sugar and 4 tablespoons water. Bring this mixture to the boil, stirring to dissolve the sugar. Add the clementine segments and lower the heat. Simmer for 5 minutes, then remove from the heat and set aside. Brush a pancake pan or 15 cm (6 in) omelette pan with kitchen paper dipped in oil. Heat over medium heat until a drop of water sputters on contact. Pour 3 to 4 tablespoons of batter into the pan and tilt the pan to coat the surface evenly with the batter.

When the pancake is browned and can be shaken loose (after 3 to 4 minutes), turn it with your fingers or a fish slice and brown the other side (2 to 3 minutes more). Slide the finished pancake on to kitchen paper to cool. If the batter is too thick, add a little milk to attain the proper consistency. Stack the pancakes as they cool. When ready to serve, spread a couple of tablespoons of the cheese across each pancake. Roll them up and place 3 on each of 4 serving plates. Spoon the clementines with their sauce over each serving.

TWO
SOUPS, SALADS
AND STARTERS

GRILLED PINK GRAPEFRUIT AND PORK SKEWERS

1 whole pork fillet, about 350 g (12 oz)

4 tablespoons soy sauce

60 g (2 oz) dark soft brown sugar

4 tablespoons freshly squeezed lemon juice

2 tablespoons dry sherry

1 tablespoon grated fresh root ginger

1 garlic clove, very finely chopped

1 pink grapefruit

Cut the pork into 2.5 cm (1 in) cubes and place them in a bowl. Combine the remaining ingredients, except the grapefruit, and pour them over the pork. Cover and marinate for 1 hour. Slice the unpeeled grapefruit into 12 sections and halve each section. Thread the pork cubes alternately with the grapefruit pieces (piercing them through the skin) on to 8 skewers. Reserve the marinade. Grill or barbecue over hot coals, turning and basting often with the reserved marinade, for about 15 minutes, or until cooked thoroughly.

Serves 4

Preparation time: 1½ hours

LIME SOUP

2 teaspoons grated lime zest

125 ml (4 fl oz) freshly squeezed lime juice

1 fresh green chilli, seeded and coarsely chopped

1 large cucumber, peeled, seeded and coarsely chopped

300 g (10 oz) seedless white grapes

75 g (2$^{1}/_{2}$ oz) coarsely chopped spring onions

4 tablespoons fresh coriander leaves

75 g (2$^{1}/_{2}$ oz) coarsely chopped green pepper

4 tablespoons golden tequila

thin slices of lime to garnish

Serves 4

Preparation time: 20 minutes

Purée all the ingredients, except the last 2, in 2 batches, using half the lime juice in each batch, in a food processor or blender until smooth. Chill thoroughly. Add the tequila, mix well and serve in chilled bowls or cups. Garnish with lime slices.

LEMON AND PARSLEY SOUP

30 g (1 oz) unsalted butter

4 bunches flat-leaf parsley, stalks removed (about 175 g/6 oz leaves)

1 teaspoon ground coriander

2 garlic cloves, crushed and peeled

750 ml (1¼ pints) rich chicken stock

4 tablespoons freshly squeezed lemon juice

1 tablespoon grated lemon zest

500 ml (16 fl oz) plain yogurt or soured cream

salt and freshly ground black pepper

parsley leaves to garnish (optional)

lemon zest to garnish (optional)

Serves 4

Preparation time: 25 minutes

Heat the butter in a large non-reactive saucepan over medium heat. Add the parsley leaves and cook, tossing frequently, until the leaves are completely wilted and coated with butter (about 3 to 5 minutes). Add the coriander and garlic and sauté for 2 minutes longer. Add the stock, lemon juice and lemon zest. Bring the mixture to the boil, then lower the heat and simmer for 10 minutes. Remove the soup from the heat and purée in batches. Wipe out the pan and return the soup to it. Add the yogurt or soured cream and mix thoroughly.

If you are serving the soup cold, allow it to cool, then cover and refrigerate it until it is well chilled. If you are serving it hot, heat the soup through, but do not let it boil. When serving, season with salt and pepper to taste and garnish with parsley and lemon zest, if desired.

CURRIED ORANGE SOUP

15 g (¹/₂ oz) unsalted butter

1 medium onion, chopped 450 g (1 lb) carrots, peeled and chopped

1 large garlic clove, peeled and crushed

750 ml (1¹/₄ pints) chicken stock

¹/₂ teaspoon ground turmeric

3 whole cloves

1 teaspoon coriander seeds

12 black peppercorns

¹/₂ teaspoon cumin seeds

1 dried red chilli (the larger the chilli, the hotter the soup)

¹/₂ teaspoon fennel seeds

1 x 7.5 cm (3 in) cinnamon stick, broken in half

6 allspice berries

2 cardamom pods

4 slices fresh root ginger, peeled, about 3 mm (¹/₈ in) thick

500 ml (16 fl oz) freshly squeezed orange juice

mint leaves to garnish

orange slices to garnish

Serves 6

Preparation time: 50 minutes

Melt the butter over medium heat in a large non-reactive saucepan. Add the onion and sauté until the edges are golden (about 10 minutes). Add the carrots, garlic, stock and turmeric. In a piece of muslin, tie the cloves, coriander seeds, peppercorns, cumin seeds, dried red chilli, fennel seeds, cinnamon stick, allspice berries, cardamom pods and ginger. Add the spice bag to the pan and bring to the boil. Lower the heat and simmer, uncovered, for 30 minutes, stirring occasionally.

Remove the spice bag and purée the soup in 2 batches in a food processor or blender. Wipe out the pan and return the puréed soup to it. Add the orange juice and heat through.

Serve hot, garnished with mint leaves and orange slices. (This soup may be made a day in advance if desired and reheated.)

SALAD OF BEETROOT, LAMB'S LETTUCE AND CLEMENTINES

4 beetroots, each 5 cm (2 in) in diameter

4 handfuls Lamb's lettuce, washed and roots trimmed

4 clementines, peeled, segmented and pith removed

2 spring onions, thinly sliced

Vinaigrette

1 tablespoon balsamic vinegar

1 tablespoon lemon juice

1 teaspoon Dijon mustard

salt and freshly ground black pepper

4 tablespoons good quality virgin olive oil

Serves 4

Preparation time: 45 minutes

Top and tail the beetroots, leaving 5 cm (2 in) of stalk and root attached. Drop them into boiling salted water and boil for 25 to 30 minutes, or until just tender. Then plunge them into a basin of cold water until cool. Cut off the ends and slice crossways into 3 mm ($^1/_8$ in) thick slices.

Arrange the beetroot slices and clementine segments around 4 serving plates. Pile the Lamb's lettuce in the centre of each plate. Scatter the spring onions over each salad. In a small bowl, combine the vinegar, lemon juice, mustard, and salt and pepper to taste. Whisk, adding the oil in a stream. When all the oil has been incorporated, drizzle the vinaigrette over the salads and serve.

UGLI FRUIT CARIBBEAN FISH SALAD

700 g (1¹/₂ lb) fillet of cod (or other mild white fish)

5 tablespoons freshly squeezed lemon juice

1 medium green pepper, seeded and cut into 2.5 cm (1 in) squares

3 medium ripe tomatoes, cut into 1 cm (¹/₂ in) wedges

1 large cucumber, peeled, seeded, and sliced 5 mm (¹/₄ in) thick

2 bananas, peeled and sliced 5 mm (¹/₄ in) thick

1 large ugli fruit, peeled, pith removed, segmented and pipped

45 g (1¹/₂ oz) freshly grated coconut

lettuce leaves (optional)

Dressing

2 tablespoons freshly squeezed lime juice

250 ml (8 fl oz) plain yogurt or soured cream

1 garlic clove, very finely chopped

1 fresh green chilli, seeded and very finely chopped

¹/₄ teaspoon ground cumin

¹/₂ teaspoon freshly ground

black pepper

salt to taste

Serves 6 to 8

Preparation time: 1 hour

Preheat oven to 220ºC (425ºF or gas 7).

Place the fish in a large non-reactive baking tin, slightly overlapping the pieces if necessary. Pour the lemon juice over the fish. Cover the tin with foil and bake for 8 to 10 minutes, or until the fish is just cooked through. Uncover the tin and allow the fish to cool.

Remove the fillets to a serving bowl. Add the remaining salad ingredients. Toss gently and set aside. The fillets will break easily into bite-size pieces.

In a small bowl, combine the dressing ingredients and mix thoroughly. Serve salad on a bed of lettuce, if liked; top with dressing.

CHICKEN SALAD WITH TANGERINES AND RED ONIONS

3 tangerines or mandarin oranges

300 g (10 oz) cooked chicken meat, coarsely chopped

60 g (2 oz) celery, chopped

1 small red onion, chopped

4 tablespoons chopped parsley

Dressing

1 x 90 g (3 oz) packet full fat soft cheese, at room temperature

125 ml (4 fl oz) mayonnaise

1 tablespoon prepared horseradish

4 tablespoons milk

salt and freshly ground black pepper

Serves 4

Preparation time: 30 minutes

Peel the tangerines. Scrape away the pith and finely chop enough peel to make 2 tablespoons. Reserve.

Separate the tangerine segments and remove any pith or pips. Halve each segment and place in a bowl with the chicken, celery, onion and parsley. To make the dressing, combine all the ingredients in the bowl of a food processor or blender, and pulse on and off for a few seconds until thoroughly combined.

Gently toss the chicken mixture with the dressing. Chill until ready to serve. Serve on any kind of lettuce or whole-grain bread. Garnish with reserved peel.

PICKLED CITRUS PRAWNS

5 tablespoons extra virgin olive oil

5 tablespoons freshly squeezed lemon juice

5 tablespoons freshly squeezed lime juice

3 tablespoons orange blossom honey

2 tablespoons capers, drained

1 teaspoon celery seed

1 tablespoon prepared horseradish

1 teaspoon Tabasco sauce

$^1/_2$ teaspoon salt

450 g (1 lb) raw king prawns in shell, peeled and deveined

1 large orange

1 small grapefruit

1 small red onion, halved and thinly sliced

lettuce leaves or rocket (optional)

Serves 6 as a starter

Preparation time: 20 minutes to assemble, plus 1 day to marinate

Combine the first nine ingredients in a large, non-reactive bowl. Whisk thoroughly to combine. Blanch the prawns for 1 minute in boiling water. Drain the prawns, add them to the bowl and toss to coat.

Peel the orange and the grapefruit and scrape off the white pith. Separate the segments, making a small slit in any segments that contain pips and popping them out. Add the segments to the bowl along with the onion and toss to coat. Cover and refrigerate overnight, tossing occasionally.

Serve on beds of lettuce leaves or rocket if liked. Drizzle over some of the marinade.

THREE

MAIN
COURSES

LEMON FETTUCCINE WITH PEPPERED PRAWNS

450 g (1 lb) fresh Lemon Fettuccine (recipe follows)

1 tablespoon salt

150 ml (¼ pint) olive oil

60 g (2 oz) unsalted butter

1 large garlic clove, very finely chopped

450 g (1 lb) raw king prawns in shell, peeled and deveined

1 large red pepper, finely chopped 1 small fresh green chilli, finely chopped

20 g (½ oz) parsley leaves, chopped

salt and freshly ground black pepper

lemon slices to garnish

Bring a large pan of water to the boil. Add 1 tablespoon of salt, 2 tablespoons of oil and the fettuccine to the water. Stir the fettuccine gently until the water returns to the boil. Boil for 1½ to 2 minutes, or until *al dente*. Drain.

Meanwhile, as the water heats to boil the pasta, put the olive oil and butter in a large frying pan and heat over medium high heat. When the fettuccine returns to the boil, add the garlic to the frying pan and sauté for a few seconds. Add the prawns and sauté for 1 minute on each side. Add the red pepper and chilli and sauté for 2 more minutes, or until prawns are opaque. Remove from the heat and add the parsley.

Arrange the fettuccine on plates. Divide the prawns and place them on the pasta. Spoon the sauce over each serving and add salt to taste and a generous grinding of black pepper. Garnish with the lemon slices.

LEMON FETTUCCINE

175 g (6 oz) fine semolina

150 g (5 oz) unbleached plain flour

3 tablespoons finely grated

lemon zest

¼ teaspoon salt

1 size-3 egg

1 size-3 egg yolk

1 tablespoon olive oil

2 tablespoons strained lemon juice

If you are using an electric pasta machine, follow the manufacturer's instructions. If you are using a manual machine, place all the ingredients in a food processor. Process until well blended. If the mixture is too dry, add additional lemon juice 1 teaspoon at a time, processing for 5 seconds after each addition. Transfer the mixture to your work surface and knead for a few seconds until it is fairly smooth, dusting it with flour if it sticks. Wrap the pasta dough tightly and set it aside for 30 minutes. Process through the machine following the manufacturer's directions. Yield: 450 g (1 lb).

Serves 4

Preparation time: 25 minutes, plus 30 minutes for pasta to rest

FRIED SCALLOPS WITH LIME AND GARLIC

3 tablespoons freshly squeezed lime juice

1 tablespoon grated lime zest

2 teaspoons Dijon mustard

2 tablespoons chopped fresh coriander

2 garlic cloves, peeled and crushed

250 ml (8 fl oz) extra virgin olive oil

700 g (1½ lb) large shelled scallops, without corals

vegetable oil for frying

120 g (4 oz) unbleached plain flour

2 tablespoons chilli powder

salt and freshly ground black pepper

2 size-3 eggs, lightly beaten

Serves 4 to 6

Preparation time: 20 minutes

Combine the lime juice and zest, mustard, coriander and garlic in the bowl of a food processor or blender. Process for a few seconds to combine; then, with motor running, add the olive oil in a stream and process until thickened and incorporated. Refrigerate until ready to serve.

Wash the scallops and lay on kitchen paper to dry. Heat 1 cm (½ in) of vegetable oil in a large frying pan over medium high heat.

In a small bowl, combine the flour, chilli powder, and salt and pepper to taste. Dredge the scallops in the mixture, rolling to coat. Shake off any excess and dip in eggs to coat. Return to flour mixture and roll until coated. Again, shake off excess. Fry the scallops in several batches, being careful not to crowd them. Turn them over as they become brown and crisp. When the second side is done, remove them and drain on kitchen paper. Don't overcook. Serve hot with individual bowls of the lime sauce for dipping.

GRILLED SHARK STEAKS WITH CITRUS SALSA

Citrus Salsa

2 medium oranges

1 small red onion, chopped

1 garlic clove, peeled and crushed

1 to 2 fresh green chillies, roasted, peeled, seeded and coarsely chopped

1 tablespoon olive oil

3 tablespoons freshly squeezed lime juice

2 tablespoons chopped coriander leaves

salt and freshly ground black pepper

Shark Steaks

4 x 9.5 cm (3³/4 in) thick shark, tuna or swordfish steaks, 175 to 225 g (6 to 8 oz) each

4 teaspoons olive oil

salt and freshly ground black pepper

Serves 4

Preparation time: 45 minutes, plus 1 hour to rest

Peel the oranges over a small bowl with a vegetable knife, cutting through the peel and just barely into the flesh, then spiralling down and around from one end to the other, as if peeling an apple. Cut into the flesh next to each dividing membrane and pop out each segment, removing any pips. Reserve. In the bowl of a food processor or blender, process the remaining salsa ingredients, pulsing on and off and scraping down the sides of the bowl until well chopped. Add the oranges and continue to process until you have a chunky, well-combined mixture. Cover and set aside at room temperature for 1 hour. Brush the shark steaks on both sides with oil, and sprinkle with salt and pepper to taste. Grill under a medium high heat for 4 to 5 minutes on each side, or until the fish feels firm but bouncy. Serve immediately with the salsa.

BAKED FISH
WITH UGLI FRUIT

30 g (1 oz) unsalted butter, softened

450 g (1 lb) white fish fillets, such as plaice

1 x 2 cm (3/4 in) piece fresh root ginger, peeled

1 small onion, thinly sliced

1 medium ugli fruit

2 teaspoons chopped fresh thyme leaves (or $^3/_4$ teaspoon dried)

salt and freshly ground black pepper

Preheat oven to 230ºC (450ºF or gas 8) and place a rack in the upper third of the oven. Grease a shallow, non-reactive tin just large enough to hold the fillets with one-quarter of the butter.

Arrange the fish in a single layer in the tin. Slice the ginger paper thin and scatter it, along with the onion slices, over thefish.

Peel the ugli fruit, removing the white pith. Separate the segments and scatter them in the tin. Top with the thyme, and salt and pepper to taste. Dot with the remaining butter and bake for about 10 minutes, or until the fish is opaque and flaky. Pour the pan juices over the fillets when served.

Serves 4

Preparation time: 20 minutes

GRAPEFRUIT-MARINATED POUSSINS

3 poussins, halved and backbones removed

75 g (2¹/₂ oz) mint jelly

250 ml (8 fl oz) freshly squeezed grapefruit juice

¹/₂ teaspoon freshly ground black pepper

¹/₂ teaspoon ground cinnamon

1 tablespoon Dijon mustard

1 garlic clove, very finely chopped

grapefruit slices to garnish

mint leaves to garnish

Serves 6

Preparation time: 1 hour, plus 4 hours marinating time

Place the poussin halves on the work surface, bone side down. With the heel of your hand, press on each breast, flattening it somewhat. Tuck the tip of the wing under the breast. With a sharp knife, make a small slit in the skin between the breast and thigh. Stick the end of the leg through the slit. Then place the pieces in a non-reactive tin or bowl just large enough to hold them.

In a small bowl, whisk together the remaining ingredients, except the grapefruit slices and mint leaves, until well combined. Pour the marinade over the poussins, cover and refrigerate for about 4 hours.

To barbecue over charcoal, cook the birds skin side down for 15 to 20 minutes. Turn and cook for another 15 minutes, until the birds are tender, basting occasionally.

To grill, put the poussins under moderately high heat and cook for 15 to 20 minutes on each side, basting occasionally with the marinade.

Garnish with grapefruit slices and mint leaves before serving.

CHICKEN WITH LEMON AND OLIVES

4 tablespoons olive oil

700 g (1½ lb) skinless, boneless chicken breasts (or thighs)

flour for dredging

2 medium onions, chopped

2 garlic cloves, chopped

1½ teaspoons

chopped fresh rosemary leaves (or ½ teaspoon dried)

2 teaspoons grated lemon zest

5 tablespoons freshly squeezed lemon juice

125 ml (4 fl oz) dry white wine or chicken stock

120 g (4 oz) pimiento-stuffed olives, drained and rinsed

freshly ground black pepper to taste

Serves 4 to 6

Preparation time: 35 minutes

Heat the oil in a large frying pan over medium-high heat. Dredge the chicken pieces in the flour, shaking off the excess. Brown the chicken in the oil, about 4 to 5 minutes on each side. Remove the chicken to a plate and set aside.

Sauté the onion and garlic in the remaining oil until translucent. Add the remaining ingredients and bring to the boil, scraping up any browned bits from the pan. When the sauce begins to boil, return the chicken to the pan and work the pieces down into the sauce, spooning some over each piece. Lower the heat to medium and cook for 2 minutes on each side for breasts or 4 to 5 minutes for thighs.

Serve over pasta or rice, if desired.

SOBA WITH FENNEL, CHÈVRE AND KUMQUATS

1 medium fennel bulb, with leaves

150 g (5 oz) kumquats, well washed

4 tablespoons olive oil

1 medium onion, thinly sliced

1 garlic clove, finely chopped

1 teaspoon chopped fresh thyme leaves (or ½ teaspoon dried)

250 ml (8 fl oz) dry white wine or chicken stock

salt and freshly ground black pepper

225 g (8 oz) soba (Japanese buckwheat noodles)

120 g (4 oz) goat's cheese

Bring a large pan of water to the boil for the soba. Meanwhile, trim the fennel bulb, reserving enough of the feathery leaves to make 2 tablespoons, chopped. Quarter the bulb and cut out most of the core. Thinly slice each piece crossways.

Thinly slice the kumquats and remove the pips. Heat the oil in a frying pan over medium high heat. Add the onion, garlic and sliced fennel and sauté for 2 to 3 minutes. Add the thyme, kumquats and wine or stock. Cover, lower the heat to medium and cook for 5 minutes. Remove from the heat and add salt and pepper to taste. Boil the soba until it is al dente. Drain and arrange on 4 plates. Spoon the sauce over the noodles. Top with crumbled goat's cheese and the reserved chopped fennel leaves.

Serves 4

Preparation time: 30 minutes

SPARERIBS WITH LIME AND CHILLI

1.3 to 1.8 kg (4 to 5 lb) pork spareribs, trimmed of excess fat

175 ml (6 fl oz) freshly squeezed lime juice

1 or 2 large fresh green chillies, seeded and coarsely chopped

4 garlic cloves, peeled and crushed

1 1/2 teaspoons fresh marjoram leaves (or 1/2 teaspoon dried)

4 tablespoons vegetable oil

Serves 4

Preparation time: 30 minutes, plus marinating time, plus 2 hours

Place the spareribs in a shallow non-reactive dish that's just large enough to hold them. Put the other ingredients in a blender or food processor and purée. Pour the marinade over the ribs. Cover and marinate for several hours or overnight, turning the ribs once or twice.

Preheat the oven to 180ºC (350ºF or gas 4).

Place the ribs on a rack in a shallow roasting tin and bake for 1 1/2 to 2 hours, basting with the marinade every 20 minutes. Turn the ribs after the first hour. When they're thoroughly browned and crisp, cut them apart, if necessary, and serve.

PORK MEDALLIONS WITH MANDARIN ORANGES AND CRANBERRIES

3 mandarin oranges

4 loin pork chops, 4 cm (1¹⁄₂ in) thick, boned

30 g (1 oz) unsalted butter

2 tablespoons olive oil

salt and freshly ground black pepper

2 garlic cloves, very finely chopped

1¹⁄₂ teaspoons chopped fresh rosemary leaves (or ¹⁄₂ teaspoon dried)

¹⁄₂ teaspoon ground cinnamon

pinch ground cloves

2 tablespoons soft brown sugar

125 ml (4 fl oz) Marsala

60 g (2 oz) cranberries

rosemary sprigs to garnish

Serves 4

Preparation time: 45 minutes

Peel the oranges. Scrape the pith from the peel and chop enough peel to make 2 tablespoons. Reserve. Remove any pith from the orange segments; separate the segments and remove any pips. Reserve.

Preheat oven to 95°C (200°F or gas 4).

Pound chops on both sides with a mallet or the side of a cleaver. Re-form the chops and tie into rounds using kitchen string.

Heat half of the butter with the olive oil in a frying pan over medium high heat. Season pork with salt and pepper to taste and sauté the medallions for 10 to 12 minutes on each side, until well browned and cooked through. Remove the strings, place the medallions on a warm serving plate and keep them warm in the oven.

Melt the remaining butter in the pan and add the garlic, rosemary, cinnamon, cloves, brown sugar and reserved chopped peel. Sauté, stirring, for a few seconds, then add the Marsala and bring to the boil. Boil until the sauce is reduced and thickened. Reduce heat to medium. Add the cranberries and cover the pan for 1 to 2 minutes, until the cranberries have burst. Uncover the pan and add the orange segments and salt and pepper to taste. Heat thoroughly and spoon the sauce over the medallions. Garnish with rosemary sprigs and serve.

VEAL CHOPS WITH BLOOD ORANGES

120 g (4 oz) unsalted butter

4 veal loin chops, about 4 cm (1½ in) thick flour for dredging

2 garlic cloves, very finely chopped

1 tablespoon chopped fresh marjoram leaves (or 1 teaspoon dried)

350 ml (12 fl oz) blood orange juice (about 7 or 8 oranges) sprigs of marjoram to garnish

orange zest to garnish

Melt three-quarters of the butter in a large frying pan over medium high heat.

Dredge the chops in flour, shaking off the excess, and place in the pan. Brown for 5 minutes, turn over and brown 5 minutes more.

Lower the heat to medium low and pour off the fat. Add the garlic, chopped marjoram and juice. Cover and cook for about 5 minutes. Uncover, turn the chops over, then recover and cook 5 minutes more. Remove the chops from the pan and keep them warm. Raise the heat, add the remaining butter and stir until it melts and is incorporated into the sauce. Pour over the chops and serve, garnished with marjoram sprigs and orange zest.

Serves 4

Preparation time: 25 minutes

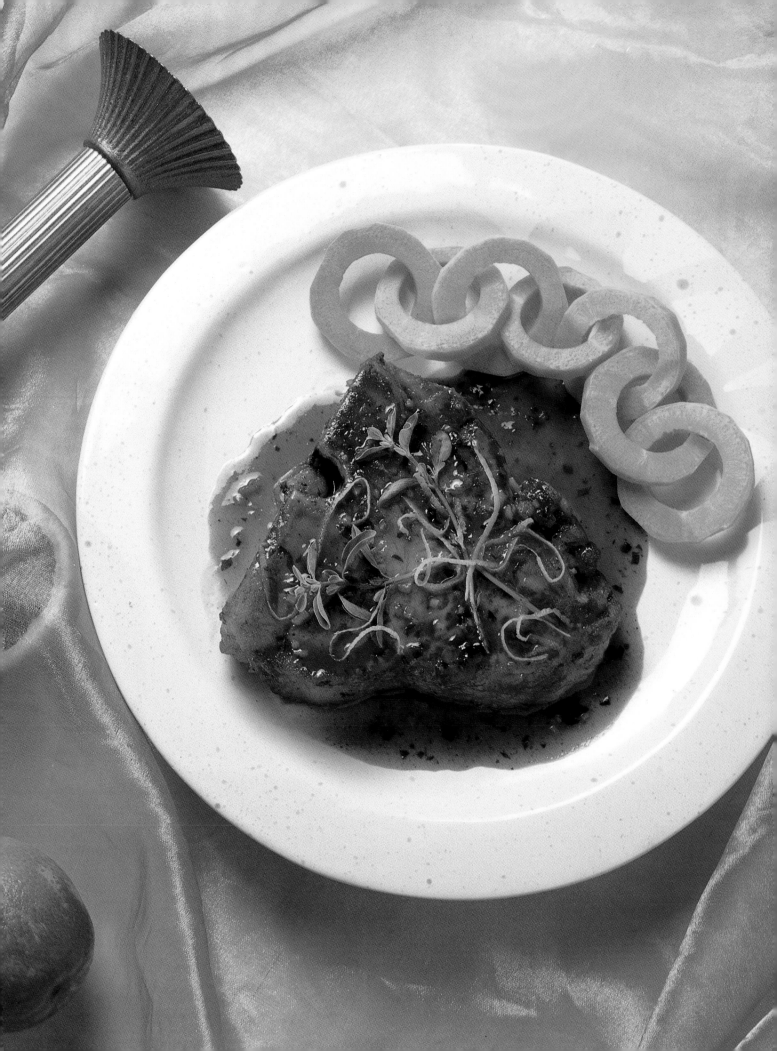

BEEF WITH TANGERINES

900 g (2 lb) lean beef chuck steak or boned topside, cut into 3 cm (1¼ in) cubes

flour for dredging

3 tablespoons vegetable oil

500 ml (16 fl oz) dry red wine

250 ml (8 fl oz) rich beef stock

350 g (12 oz) button onions, peeled

2 large carrots, peeled and cut into 2.5 cm (1 in) chunks

2 large garlic cloves, peeled and crushed

3 tablespoons soy sauce

½-¾ teaspoon dried chilli flakes, to taste

1 teaspoon caster sugar

5 slices fresh root ginger, peeled, about 3 mm (⅛ in) thick

2 tangerines or mandarin oranges

1 large green pepper, seeded and cut into 2.5 cm (1 in) squares

1 large red pepper, seeded and cut into 2.5 cm (1 in) squares

30 g (1 oz) thinly sliced spring onion greens

cooked rice (optional)

Serves 6

Preparation time: 2¾ hours

Dredge the beef cubes in flour, shaking off any excess. Brown the meat in batches in the oil over high heat in a flameproof casserole. Add more oil if needed.

Return meat to the casserole; add wine, stock, onions, carrots, garlic, soy sauce, chilli flakes, sugar and ginger. Bring to the boil, then lower heat, cover and simmer for 1 hour.

Near the end of the simmering hour, remove the skin from the tangerines in as large pieces as possible, reserving the fruit. Lay the peel, outer side down, on a hard surface and, using the edge of a sharp knife or grapefruit spoon, scrape off as much white pith as possible. Cut the scraped peel into 1 cm (½ in) squares and add them to the casserole. Simmer for 30 minutes longer. Add the peppers and simmer for another 15 minutes.

Separate the reserved tangerine segments and peel off as much white pith as possible. Make a small slit in the inner edge of any segments with pips and pop the pips out through the slit. Add the tangerine segments to the casserole and heat through for 5 minutes.

Serve, sprinkled with spring onion greens, over rice if liked.

FOUR
RELISHES AND CONDIMENTS

PICKLED LEMONS

10-12 lemons

6 garlic cloves, crushed

6 slices fresh root ginger, peeled, about 3 mm (^{1}/$_8$ in) thick

1 teaspoon ground turmeric

2 teaspoons cumin seeds

2 tablespoons yellow mustard seeds

1-2 teaspoons dried chilli flakes, to taste

^{1}/$_2$ teaspoon fennel seeds

1 teaspoon fenugreek seeds

1 litre (1^3/$_4$ pints) distilled white vinegar

3 tablespoons salt

Makes 2 litres (3^1/$_2$ pints)
Preparation time: 30 minutes to assemble, plus 3 weeks to pickle

Scrub the lemons well with a brush, using warm soapy water. Rinse thoroughly. Quarter them and place them, interspersed with the garlic cloves and ginger slices, in a 2 litre (3^1/$_2$ pint) wide-mouthed jar, leaving 2.5 to 5 cm (1 to 2 in) of space at the top of the jar.

Heat a frying pan over medium low heat and add the turmeric, cumin seeds, mustard seeds, dried chilli flakes, fennel seeds and fenugreek seeds. Heat the mixture carefully, shaking the pan frequently, until you begin to smell the spices and they begin to smoke. Remove them from the heat and add 250 ml (8 fl oz) of vinegar to the pan. Mix thoroughly and carefully pour the vinegar and spices over the lemons in the jar. Add the salt and enough additional vinegar to cover the lemons. Seal the jar, shake it, and roll it around to mix the ingredients. Leave in a cool dark place for 3 weeks, shaking the jar and rolling it around 2 or 3 times a week. When 2 or 3 weeks have passed, chill the lemons. They will keep in the refrigerator for 2 to 3 months. Slice and serve as a pickle, or chop and use as a condiment with curries, meats or salads.

CITRUS MARMALADE

1 small unblemished grapefruit

1 large unblemished orange

1 large unblemished lemon

1 large unblemished lime

800 g (1³/₂ lb) preserving sugar

2 tablespoons whisky (optional)

Makes about 1.5 litres (2¹/₂ pints)

Preparation time: 2¹/₃ hours over 2 days, plus 2 or 3 weeks to stand

Scrub the fruit well with a brush, using warm soapy water if they're not organic. Rinse thoroughly. Slice some peel from each end of each fruit, just deep enough to reach the fruit. With a sharp knife, cut the grapefruit and orange into eighths and the lemon and lime into quarters lengthways. Lay the wedges on a flat surface with one of the cut sides down and the peel side towards you. Slice the wedges as thinly as possible so that you end up with little triangles. Try to keep as much of the juice as possible in the fruit.

Place the fruit slices in a preserving pan with 2 litres (3¹/₂ pints) of water. Bring to the boil and boil rapidly for 35 to 40 minutes, or until the peel is tender, occasionally stirring down any pulp pushed up the sides of the pan. Cover and leave overnight.

The following day, add the sugar to the pan and heat over medium low heat, stirring until the sugar is dissolved. Raise the heat and boil rapidly, stirring frequently, for 15 to 20 minutes, or until thickened, so that the syrup sets after cooling several minutes on the spoon or after 1 minute in the freezer. Be careful that the mixture does not stick to the pan. Remove from heat and stir in whisky, if desired. Leave the marmalade for 5 minutes, allowing it to thicken enough to maintain an even distribution of the pieces of peel. Skim off any foam, stir and spoon into sterilised jars. Seal and process in a boiling-water bath (directions follow) for 10 minutes, or cool and store in the refrigerator for 2 to 3 weeks.

BOILING-WATER BATH

A boiling-water bath is not suitable for bottling all types of food, but it works well for finishing and sealing the preserved foods in this chapter. Its purpose is to kill yeasts, moulds and bacteria that cannot live at 100°C (212ºF), the temperature of boiling water. It also forces out any air trapped in the tissues of the food and in the jar itself. Thus, it creates a vacuum that allows the jars to seal themselves for longer storage.

I find it easiest to use 250 ml (8 fl oz) jars because the size of the pan increases with the jar's size. For these jars you will need a 14.5 litre (6¹/₂ quart) pan and a rack, which will allow water to circulate and hold the jars 2 to 2.5 cm (³/₄ to 1 in) above the bottom of the pan. The same pan can be used to sterilise the lids.

Place the jars on the rack with at least 2.5 cm (1 in) between them. Fill the pan with water to 4 cm (1¹/₂ in) above jars.

Start timing and boil the jars for 10 minutes. Remove jars with tongs and turn off heat. Fill jars with the hot preserve. Seal with new lids and return them to the pan of hot water with at least 2.5 cm (1 in) between them. They should be covered by 2.5 to 5 cm (1 to 2 in) of water. Add more boiling water if necessary. There should he a 2.5 to 5 cm (1 to 2 in) space between the water and the lid for `boiling room'. Turn heat back on high. Bring to a full rolling boil. Then start timing. Process for 10 to 15 minutes at a full boil.

Remove jars with tongs and allow to cool.

KUMQUAT AND GINGER PRESERVE

900 g (2 lb) kumquats

1 tablespoon bicarbonate of soda

600 g (1¼ lb) preserving sugar

1½ tablespoons fresh root ginger, peeled and grated

Makes 1 litre (1¾ pints)

Preparation time: 2½ hours, over 2 days

Wash the kumquats thoroughly in warm soapy water. Remove the stalk ends with your fingernail or a small knife. Place the kumquats in a preserving pan. Sprinkle with bicarbonate of soda and cover with boiling water. Allow to soak for 10 minutes, then drain and rinse the pan and the kumquats very thoroughly.

Using a sharp knife, cut an 'X' about 1 cm (½ in) deep on the stalk end of each kumquat. Return kumquats to the pan. Add fresh water to a level about 2.5 cm (1 in) above the fruit. Boil the fruit for 15 minutes over high heat, then drain and return to the pan. Add more fresh water and boil for another 15 minutes. Then drain, and repeat the process a third time. Drain again and set aside.

Place the sugar in the pan with 750 ml (1¼ pints) of fresh water. Bring to the boil and cook at a full boil for 5 minutes. Add the kumquats and the ginger. Return to the boil and boil for about 30 minutes or until the kumquats are becoming translucent. (Keep the heat as high as possible without boiling over. Bubbles will rise to the top of the pan, so don't leave it unattended. Scrape any ginger that adheres to the side of the pan back into the liquid.)

Remove the pan from the heat, cover and leave overnight so the kumquats can plump up.

The following day, return the pan to the boil and boil for about 5 minutes, until the syrup is as thick as honey and a small amount of syrup in a spoon sets after 1 minute in the freezer.

Pack the kumquats into sterilised jars and pour the hot syrup over them, leaving 1 cm (½ in) of head space. Wipe the rims clean and seal the jars.

If you plan to use the preserves within 2 weeks, cool and refrigerate. If you'll be storing them longer, process in a boiling-water bath (see note on page 80) for 10 minutes and store in a cool dark place.

LIME AND TOMATO RELISH

2 large thin-skinned limes, thinly sliced and pipped, with each slice quartered

4 large ripe tomatoes, cored and chopped

1 large onion, peeled and chopped

1 medium green pepper, seeded and chopped

1 fresh chilli, finely chopped

1 garlic clove, finely chopped

$1^1/_2$ teaspoons mustard seeds

$^1/_2$ teaspoon celery seeds

$^1/_4$ teaspoon ground cloves

1 teaspoon salt

75g ($2^1/_2$ oz) preserving sugar

250ml (8 fl oz) white vinegar

Makes 2 litres ($3^1/_2$ pints)

Preparation time: 45 minutes, plus 1or 2 weeks to stand

Combine all the ingredients in a preserving pan. Bring to a full boil and remove from the heat. Seal in hot sterilised jars. If storing, process in a boiling-water bath (see note on page 80) for 10 minutes and store in a cool dark place.

Tastes best if allowed to stand a week or two before serving. Serve with seafood or grilled meats.

TANGELO AND ONION RELISH

2 large unblemished tangelos

2 large Spanish onions, peeled and chopped

$1^1/_2$ teaspoons fennel seeds

$^1/_2$ teaspoon dried chilli flakes (optional)

$^1/_2$ teaspoon ground cloves

$^1/_4$ teaspoon ground turmeric

$^1/_2$ teaspoon ground white pepper

$^1/_2$ teaspoon salt

Makes about 1 litre ($1^3/_4$ pints).
Preparation time: 45 minutes.

Use a vegetable peeler, pulling towards you, to remove the zest from the tangelos in strips. Stack the strips and cut them crossways into 3 mm ($^1/_2$ in) thick strips. Reserve.

Using a vegetable knife, hold 1 tangelo over a preserving pan and cut through the white pith just into the flesh. Remove the pith in a long strip, spiralling down and around from 1 end to the other, as if peeling an apple. Cut between the membranes with the knife and pop the fruit segments out into the pan, making sure to remove any pips. Squeeze the translucent membranes over the pan to extract any remaining juice. Repeat the process with the second tangelo.

Add the onion, the spices, the reserved zest and 125 ml (4 fl oz) water to the pan and cook over medium heat until the mixture comes to the boil. Then lower the heat and simmer until the onion begins to look translucent (about 5 minutes).

Store in the refrigerator for up to 2 weeks in airtight containers. Or seal in sterilised jars and process in a boiling-water bath (see note on page 80) for 10 minutes, then store in a cool, dark place.

Serve with hamburgers, roast or grilled meats, or stir into a pilaf.

TANGERINE AND CRANBERRY CHUTNEY

4 tangerines or mandarin oranges

375 g (12 oz) fresh cranberries

400 g (14 oz) preserving sugar

125 ml (4 fl oz) distilled white vinegar

75 g (2½ oz) sultanas

2 tablespoons fresh root ginger, peeled and chopped

1 garlic clove, very finely chopped

1 fresh red chilli, seeded and finely chopped

1 x 7.5 cm (3 in) cinnamon stick

6 whole cloves

4 allspice berries

Makes 1.5 litres (2½ pints)

Preparation time: 45 minutes, plus at least 1 week for maturing

Peel the tangerines. Scrape the white pith strands from the removed peel and sliver enough to make 4 tablespoons.

Place the peel in a preserving pan with the cranberries, sugar, vinegar, sultanas, ginger, garlic and chilli. Tie the spices in a piece of muslin and add to the pan. Cook over medium heat, stirring constantly, until the sugar is dissolved and the cranberries begin to pop. Simmer the mixture for 8 to 10 minutes, stirring frequently. Meanwhile, scrape off any white pith from the tangerines. Separate the segments and cut off the inner edge of each one, removing any pips. Chop the tangerine segments coarsely and add them to the pan. Raise the heat and bring the mixture to the boil, then simmer for 5 minutes until it has thickened, stirring often to prevent sticking.

Remove the spice bag and ladle the mixture into hot sterilised jars. Seal and process in a boiling-water bath (see note on page 80) for 10 minutes. Then store in the refrigerator or a cool place for up to 3 weeks.

ORANGE CURD

2 oranges

300 g (10 oz) preserving sugar

2 size-3 eggs

3 size-3 egg yolks

pinch of salt

225 g (8 oz) unsalted butter

Makes about 500 ml (16 fl oz)

Preparation time: 30 minutes

Using a vegetable peeler, remove the zest from both oranges. Cut the zest into 1 cm (½ in) pieces and place in the bowl of a food processor or blender with one-third of the sugar. Process on and off for about 30 seconds, or until the zest is finely chopped. Reserve.

Squeeze both oranges thoroughly and reserve the juice.

Put water 5 cm (2 in) deep into the bottom of a double boiler, or a saucepan, and bring to a simmer over medium heat. With an electric mixer or a whisk, beat together the eggs, egg yolks and salt until foamy. Gradually add the remaining sugar and then the zest mixture and continue beating until very thick (about 2 minutes). Pour the mixture into the top of the double boiler, or a heatproof bowl, and place it over the simmering water. Whisk constantly while adding the reserved orange juice. Continue whisking constantly until it is thick, steaming hot and smooth, and the zest pieces no longer sink to the bottom (about 12 minutes). Add butter a few knobs at a time, whisking until it is incorporated.

Spoon the butter into hot dry sterilised jars and allow to cool. Then seal and refrigerate for up to 2 months. Spread on bread or toast or use to sandwich cakes or biscuits. Or, while hot, spoon into baked individual tart shells, allow to cool and serve chilled with fresh fruit.

FIVE
DESSERTS

MAPLE AND ORANGE AMBROSIA

4 large oranges, peeled and pith removed

5 tablespoons maple syrup

4 tablespoons dark rum

$^{1}/_{4}$ teaspoon ground cinnamon

90 g (3 oz) large walnut pieces

pomegranate seeds (optional)

Serves 4 to 6

Preparation time: 30 minutes, plus several hours to chill

Halve the oranges lengthways and cut into 5 mm ($^{1}/_{4}$ in) thick slices. Place in a serving bowl.

Put the maple syrup, rum and cinnamon in a small bowl, stirring constantly until well blended. Pour over oranges, tossing gently to coat. Cover and refrigerate for several hours or overnight. Preheat oven to 180ºC (350ºF or gas 4). Put the walnut pieces in a small shallow tin and toast them for 15 minutes or until lightly browned. Cool.

When you're ready to serve, toss the nuts with the oranges, sprinkle with pomegranate seeds if using and serve immediately.

CHOCOLATE-DIPPED CITRUS SEGMENTS

175 g (6 oz) plain or bittersweet chocolate

60 g (2 oz) unsalted butter

350 g (12 oz) mixed citrus segments from easy to peel fruits, such as navel oranges, tangerines, ugli fruit and clementines, or whole thin-skinned lemons cut into eighths

finely chopped pistachio nuts (optional)

Melt the chocolate and butter in the top of a double boiler over simmering water, stirring to mix thoroughly. Remove any white pith from the citrus segments (except lemons) as well as any pips, by making a small slit in the membranes with a sharp knife and popping the pips out.

Dip each segment into the chocolate, coating about half of its length. Place the dipped segments on parchment paper or greaseproof paper and refrigerate until the chocolate is set. If desired, sprinkle pistachios over the dipped sections while the chocolate is still moist.

Serves 6 to 8

Preparation time: 30 minutes

MINTED RUBY GRAPEFRUIT ICE

150 g (5 oz) caster sugar

500 ml (16 fl oz) ruby red grapefruit juice, freshly squeezed, with pulp (4 or 5 grapefruits)

2 tablespoons finely chopped mint leaves

mint sprigs to garnish

In a small saucepan, heat the sugar with 125 ml (4 fl oz) of water, stirring until the sugar is dissolved. Boil for 5 minutes.

Cool to room temperature, then combine the sugar syrup with the grapefruit juice and the chopped mint leaves. Freeze the mixture in an ice cream maker, following the manufacturer's instructions, or place it in a freezer-proof container and freeze for several hours or overnight. Serve garnished with mint sprigs.

Serves 6. Makes 500 ml (16 fl oz)

Preparation time: 30 minutes, plus several hours to freeze

TANGERINE SORBET

100 g (3^1/$_2$ oz) caster sugar

2 tablespoons finely chopped tangerine or mandarin orange peel, with all white pith removed

500 ml (16 fl oz) freshly squeezed tangerine juice (about 6 to 8 tangerines)

In a small non-reactive saucepan, combine the sugar, peel and 125 ml (4 fl oz) water. Bring the mixture to the boil, stirring until all the sugar has dissolved. Boil for 5 minutes. Remove from heat and allow to cool to room temperature.

Combine the syrup with the juice and freeze in an ice cream maker, following the manufacturer's instructions, or place in a freezerproof container and freeze for several hours or overnight before serving.

Serves 6. Makes 500 ml (16 fl oz)

Preparation time: 30 minutes, plus several hours to freeze

LEMON SNAPS

375 g (12$^{1}/_{2}$ oz) unbleached plain flour

300 g (10 oz) caster sugar

2 teaspoons bicarbonate of soda

$^{1}/_{4}$ teaspoon salt

2 tablespoons grated lemon zest

175 ml (6 fl oz) vegetable oil

125 ml (4 fl oz) freshly squeezed lemon juice

2 teaspoons vanilla essence

Preheat oven to 180°C (350°F or gas 4).

Place all the ingredients in the bowl of a food processor and process until well blended, or beat with an electric mixer or whisk.

Drop by the teaspoonful on to greased baking sheets about 5 cm (2 in) apart. Bake for about 10 minutes, or until the edges of the biscuits are golden brown.

Cool on racks and store in an airtight container.

Makes about 48 biscuits

Preparation time: 40 minutes

CHOCOLATE AND ORANGE MARMALADE BROWNIES

175 g (6 oz) shelled pecan nuts or walnuts

150 g (5 oz) unsalted butter

150 g (5 oz) plain chocolate

325 g (11 oz) good-quality orange marmalade

3 tablespoons whisky

1 teaspoon orange flower water

150 g (5 oz) unbleached plain flour

$^1\!/_2$ teaspoon salt

2 tablespoons cocoa powder

4 size-3 eggs

275 g (9 oz) caster sugar

Makes 18 brownies

Preparation time: 1 hour

Preheat oven to 190ºC (375ºF or gas 5).

Spread the nuts in a large shallow baking tin and toast in the oven for about 8 minutes or until lightly toasted. Remove and cool, then break them into pieces and reserve.

Grease a 33 x 23 x 4 cm (13 x 9 x 1$^1\!/_2$ in) baking tin with 1 teaspoon of the butter. In the top of a double boiler, over simmering water, place the remaining butter and the chocolate. Cover and heat until the chocolate is almost completely melted. Remove from the heat. Uncover and stir until smooth. Add the marmalade, stirring until it has melted and combined with the chocolate. Stir in the whisky and orange flower water and set aside until it is warm, but no longer hot. Sift the flour with the salt and cocoa powder and reserve.

With an electric mixer or whisk, beat the eggs until frothy. Add the sugar and beat until light coloured and thickened (about 5 minutes). With the mixer on low, add the warm chocolate mixture, scraping the bowl and beating the mixture for a few seconds to combine. Add the flour mixture. Scrape the bowl and beat just until the flour has been combined.

Do not overbeat.

By hand, stir in the nuts. Pour the batter into the prepared tin, smoothing it out with a spatula.

Bake for about 25 minutes, or until a wooden cocktail stick inserted in the centre comes out clean.

Cool in the tin. Cut into 7.5 x 5 cm (3 x 2 in) brownies and serve or wrap individually in cling film. Full flavour develops after 24 hours; serve plain or topped with ice cream.

ORANGE AND COCONUT CUSTARD

2 medium oranges

250 ml (8 fl oz) milk

3 size-3 eggs

3 size-3 egg yolks

100 g (3½ oz) caster sugar

1 x 450 g (1 lb) can coconut cream

100 g (3½ oz) desiccated coconut

additional zest to garnish (optional)

Serves 8

Preparation time: 1½ hours, set overnight

Preheat oven to 180ºC (350ºF or gas 4).

With a vegetable peeler, remove the zest from the oranges in long strips. Reserve the oranges.

Place the zest in a small heavy saucepan with the milk and heat over medium low heat until it is steaming hot. Remove from the heat. Do not let the mixture boil.

In a mixing bowl, whisk together the eggs, egg yolks and sugar, beating until frothy. Add the coconut cream, whisking to mix well.

Remove the zest from the milk and discard. Add the hot milk to the mixing bowl in a stream while whisking. Strain the mixture and reserve.

Take the reserved oranges and, with a vegetable knife, remove the white pith and cut just through the membranes and into the fruit. Spiral down and around the fruit as if peeling an apple. Insert the knife next to the dividing membranes and pop out each segment. Remove the pips and coarsely chop the fruit. Divide the fruit among eight 175 ml (6 fl oz) ramekins. Fill the ramekins ¾ full with the custard mixture and sprinkle 1 tablespoon of desiccated coconut on the top of each one. Place the ramekins in a baking tin large enough to hold them. Fill with hot water to half the depth of the ramekins. Cover the tin with foil and bake the custard for 50 to 55 minutes or until set.

Serve warm, or cool and refrigerate for a couple of hours or overnight before serving. Garnish with zest if desired.

Lime Mousse Tart

1 tablespoon powdered gelatine

3 size-3 eggs

200 g (7 oz) caster sugar

5 tablespoons freshly squeezed lime juice (about 4 limes)

2 teaspoons finely grated lime zest

2 tablespoons icing sugar

1 teaspoon cornflour

250 ml (8 fl oz) double cream, cold

pinch salt

1 x 23 cm (9 in) tart shell, baked blind and completely cooled

Serves 8

Preparation time: 2 hours, plus 4 hours for cooling

In a small bowl, soften the gelatine in 4 tablespoons cold water. Set aside. Separate the eggs, reserving the whites at room temperature. Place the yolks in the top of a double boiler over simmering water. Add two-thirds of the caster sugar and cook, stirring constantly, until thickened, about 5 minutes. Add the softened gelatine to the hot custard and stir well, dissolving the gelatine completely. Add the lime juice and zest and mix thoroughly. Cool the mixture and chill until slightly thickened, about 1 hour. Place the icing sugar and the cornflour in a small saucepan and gradually add 4 tablespoons of the cream. Bring to the boil, stirring constantly, then simmer briefly until the mixture thickens. Remove from the heat and cool to room temperature.

When the custard has cooled and thickened, whisk the egg whites with the salt until they begin to hold a soft shape. Add the remaining caster sugar 1 tablespoon at a time, whisking until the sugar is well incorporated and the whites are stiff but not dry. Using the same whisk, whip the cream in a chilled bowl until the whisk leaves marks. Add the cornflour mixture in a steady stream and continue whisking until the cream holds stiff peaks. Do not overbeat. Fold the cream into the custard and then gently fold in the egg whites. Spoon into the tart shell and refrigerate for at least 4 hours before serving.

FROZEN BLOOD ORANGE SOUFFLÉ

4 large blood oranges

4 tablespoons Grand Marnier or other orange liqueur

100 g (3$^1/_2$ oz) caster sugar

1 tablespoon powdered gelatine

3 size-3 egg whites

pinch of salt

250 ml (8 fl oz) double cream

Serves 4

Preparation time: 3$^1/_2$ to 4 hours

Cut six 7.5 cm (6 in) long pieces of greaseproof paper (make sure they are large enough to extend around the rims of six 175 ml [6 fl oz] ramekins). Fold each piece in half lengthways and fasten around the top edge of each ramekin with string or rubber bands. The paper should extend 5 to 6 cm (2 to 2$^1/_2$ in) above the rims.

Grate enough zest from the oranges to make 4 tablespoons loosely packed. Place the zest in a small stainless steel, glass or enamel pan along with the Grand Marnier and the sugar. Simmer over low heat for 10 to 15 minutes, stirring frequently, until all the sugar is dissolved and the zest is translucent. Remove from heat.

Peel the oranges with a sharp knife, carefully removing all the white pith. Cut into segments and scrape out the pulp, removing the pips in the process.

Place the pulp in the bowl of a food processor along with the zest mixture and purée briefly. Place the mixture in a stainless steel, glass or enamel pan. Sprinkle the gelatine over the mixture, and set it aside for a few minutes to soften. Then heat the mixture over medium low heat, stirring constantly, until all of the gelatine is dissolved. Remove it from the heat, put it into a large bowl and cover it. Place the bowl in the refrigerator and chill, stirring occasionally, until the mixture mounds when dropped from a spoon (about 30 minutes).

In a small mixing bowl, whisk the egg whites with the salt until they hold stiff peaks. Add the egg whites to the orange mixture. Using the same whisk, whip the cream in another bowl until it holds stiff peaks. Do not overbeat. Add the whipped cream to the other ingredients and gently fold together until thoroughly combined. Spoon the mixture into the ramekins and freeze for 2 to 3 hours, until set. Remove the paper collars before serving.

LEMON AND NUTMEG CHEESECAKE

120 g (4 oz) rich tea biscuits, crushed with a rolling pin or in a food processor

300 g (10 oz) caster sugar

3 tablespoons finely grated lemon zest

1½ teaspoons freshly grated nutmeg

60 g (2 oz) unsalted butter, melted

900 g (2 lb) full fat soft cheese, at room temperature

50 g (1½ oz) unbleached plain flour

4 size-3 eggs, at room temperature

125 ml (4 fl oz) double cream, at room temperature

125 ml (4 fl oz) freshly squeezed lemon juice

2 teaspoons vanilla essence

Serves 8 to 10

Preparation time: 2 hours, set 4 hours or overnight

Preheat oven to 180°C (350°F or gas 4).

Generously grease a 23 cm (9 in) springform cake tin and place it in the centre of a 30 cm (12 in) square of foil. Fold the foil up around the sides of the tin and press to make a tight fit.

In a small bowl, combine the biscuit crumbs with 3 tablespoons of sugar, 2 tablespoons of zest and ½ teaspoon nutmeg. Stir to blend. Add the melted butter and mix thoroughly. Press the crumb mixture over the bottom of the tin and up the sides about 2.5 cm (1 in). Bake the crust for 10 minutes. Remove it from the oven and allow to cool for 10 to 15 minutes.

With an electric mixer or whisk, beat the soft cheese until softened. Add the remaining sugar and beat until light and fluffy. Add the flour and remaining nutmeg and beat until combined. Add the eggs, one at a time, beating well after each addition.

Combine the double cream, lemon juice, vanilla essence and remaining zest and fold it into the mixture until it is smooth. Pour the filling into the crust and bake in the centre of the preheated oven for 45 minutes or until the top is golden. Turn off the heat and allow the cheesecake to sit undisturbed in the oven for 45 minutes.

Remove the cheesecake from the oven and allow it to cool completely in the tin. Cover and chill for 4 hours or overnight. Remove the sides of the tin and allow the cake to warm slightly before serving.

Lemon Steam Cake with Blood Orange Sauce

3 size-3 eggs

200 g (7 oz) caster sugar

grated zest of 2 lemons

215 g (7½ oz) unbleached plain flour

2 teaspoons baking powder

¼ teaspoon salt

250 ml (8 fl oz) double cream

1 teaspoon vanilla essence

Serves 6 to 8

Preparation time: 1¼ hours

Grease and flour a 23 cm (9 in) springform cake tin. Place it in the centre of a 30 cm (12 in) square of foil and fold the foil up the sides of the tin to make the bottom watertight.

In a large mixing bowl, beat the eggs until they are thick and pale yellow. Beat in the sugar until the mixture is smooth and well combined. Beat in the lemon zest. Sift the flour, baking powder and salt together. Combine the cream and vanilla essence.

Add the flour mixture and the cream mixture alternately to the egg mixture, about a third at a time, beating well after each addition. When it is well mixed, pour the batter into the tin. Place the tin on a rack in a wok or large pan filled with enough simmering water to almost reach the rack. (A vegetable steamer works well.) Cover tightly and steam over medium low heat for 45 to 50 minutes until a metal skewer inserted in the centre comes out clean. Remove the cake to a rack and allow to cool.

When the cake has cooled completely, run a knife around the edge of the tin and carefully remove the sides. Slice and serve with Blood Orange Sauce (recipe follows).

Blood Orange Sauce

250 ml (8 fl oz) strained blood orange juice (5-6 oranges)

100 g (3½ oz) sugar

2 whole cloves

1 teaspoon cornflour

1 tablespoon Grand Marnier or other orange liqueur

Combine the juice, sugar and cloves in a small non-reactive pan over medium heat. Stir the mixture until the sugar dissolves. Simmer for 5 minutes.

Dissolve the cornflour in the Grand Marnier and stir it into the sauce. Simmer for another 5 minutes.

Remove the cloves and serve over the cake, either hot, at room temperature or chilled.

MANDARIN ORANGE UPSIDE-DOWN CAKE

5 mandarin oranges

4 tablespoons Grand Marnier or orange liqueur

175 g (6 oz) unsalted butter, softened

75 g (2½ oz) light soft brown sugar

3 size-3 eggs, separated

1 teaspoon vanilla essence

175 g (6 oz) plain flour

150 g (5 oz) caster sugar

¼ teaspoon baking powder

¼ teaspoon bicarbonate of soda

¼ teaspoon salt

125 ml (4 fl oz) plain yogurt or soured cream

whipped cream for garnish (optional)

Serves 6 to 8

Preparation time: 2 hours

Peel the oranges and separate the segments, scraping off as much of the white pith as possible. Cut off the inner edge of each segments and remove any pips. Place the segments in a small non-reactive bowl and toss gently with the Grand Marnier. Set aside for 30 minutes, tossing occasionally, and reserve. Preheat oven to 180°C (350°F or gas 4).

Melt 60 g (2 oz) of the butter in a well-seasoned 25 cm (10 in) cast iron frying pan over medium heat. (Pan must have an ovenproof handle.) Stir in the brown sugar until it's well moistened. Spread the sugar mixture evenly over the bottom of the pan and remove it from the heat.

Arrange the marinated orange segments in a spiral pattern on the bottom of the pan, starting from the centre and working outwards. Reserve the liquid.

Add the egg yolks and vanilla essence to the liquid remaining in the bowl. Beat lightly to combine and set aside. In a large mixing bowl, combine the flour, caster sugar, baking powder, bicarbonate of soda and salt. With an electric mixer, beat on low speed for 20 seconds to blend. Add the yogurt or soured cream and the remaining softened butter and continue to beat on low speed until well combined. Increase speed to medium and beat for 2 minutes. Add the egg yolk mixture, a third at a time, beating well after each addition. Set aside.

In a small mixing bowl, whisk the egg whites until stiff. Fold a third of the egg whites into the batter. When well incorporated, gently fold in the remaining egg whites.

Pour the batter carefully over the orange segments so that you do not disturb their arrangement. Make sure the batter is distributed evenly and all orange segments are covered. Bake in the lower third of the oven for 40 to 45 minutes. The cake is done when it springs back when touched lightly in the centre or a wooden cocktail stick inserted in the centre comes out clean. Allow to cool in the pan for 5 minutes. Then run a knife around the inside of the pan, place a serving plate over the pan, invert and carefully lift the pan off the cake. Replace any orange segments that stick to the pan. Serve warm or at room temperature, with whipped cream if liked.

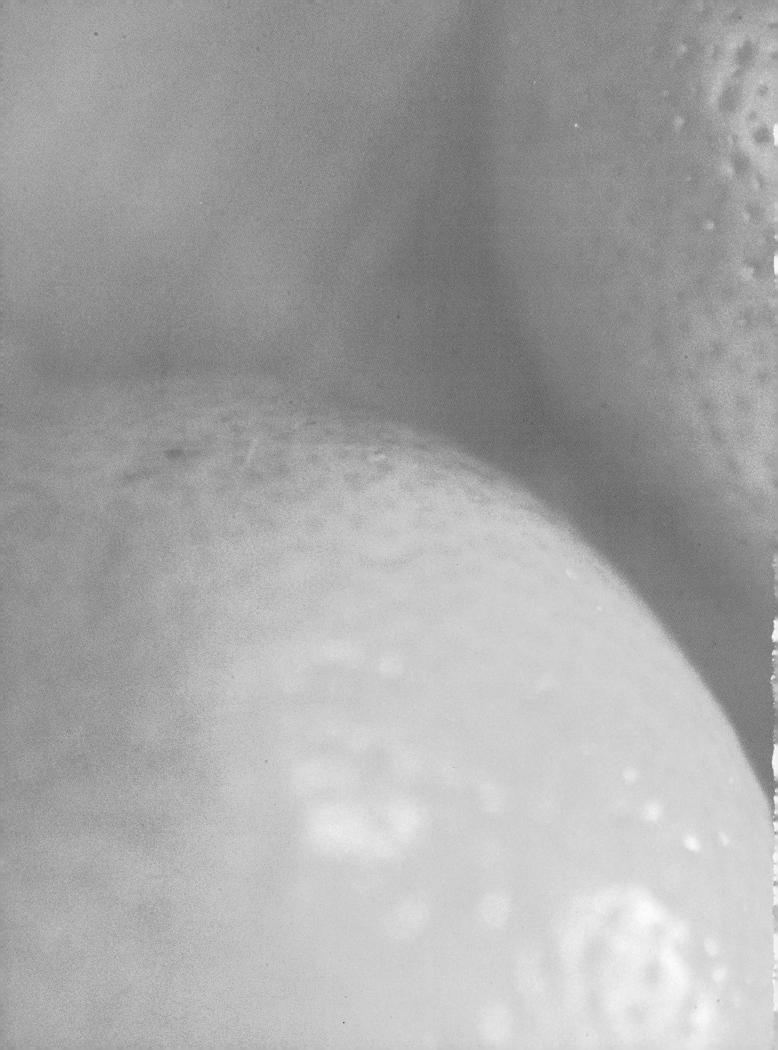

SIX
BEVERAGES

PINK LEMONADE

175 ml (6 fl oz) freshly squeezed lemon juice (4-5 lemons)

135 g (4^1/$_2$ oz) caster sugar, or to taste

^{1}/$_2$ lemon, thinly sliced and pipped

several drops of grenadine syrup

Combine the lemon juice and sugar in a 1 litre (2 pint) jug. Stir to dissolve the sugar.

Add 750 ml (1^1/$_4$ pints) of cold water and the lemon slices and mix well. Add drops of grenadine syrup until you achieve the desired 'pink-ness'.

Serve over ice.

Makes 1 litre (1^3/$_4$ pints)

Preparation time: 10 minutes

ICED ORANGE COFFEE

600 ml (1 pint) strong black coffee

1 small orange

cream and sugar

ground cinnamon

orange slices to garnish (optional)

Brew the coffee.

Using a vegetable peeler, remove the zest from the orange in strips. Put the strips in the hot coffee. Set aside for 1 hour to cool to room temperature.

Strain the coffee.

Squeeze the orange and add the juice to the coffee. Serve over ice with cream and sugar, as desired. Sprinkle the top of each serving with a pinch of cinnamon and garnish with an orange slice if desired.

Serves 4

Preparation time: 5 minutes, plus 1 hour to cool

LIMEADE FIZZ

By the glass

2 tablespoons freshly squeezed lime juice

1 tablespoon caster sugar

60 ml (2 fl oz) vodka, gin, rum or tequila

125 ml (4 fl oz) fizzy mineral water

crushed ice

lime slice to garnish

Serves 1

Preparation time: 3 minutes

In a tall glass, place the lime juice, sugar and liquor. Stir well. Add the mineral water and fill the glass with crushed ice.

Garnish with lime slice and serve.

By the jug

150 ml (^{1}/4 pint) freshly squeezed lime juice

75 g (2^1/2 oz) caster sugar

300 ml (1/2 pint) vodka, gin, rum or tequila

500 ml (16 fl oz) fizzy mineral waterlime slices

crushed ice

Serves 4

Preparation time: 5 minutes

Combine the lime juice, sugar and liquor in a 1 litre (2 pint) jug. Stir until the sugar is dissolved. Add the mineral water and the lime slices.

Stir gently to combine. Pour into glasses filled with crushed ice.

CITRUS FRAPPÉ

juice of 1 large lime

juice of 1 small grapefruit

juice of 1 medium lemon

juice of 1 medium orange

100 g (3^1/2 oz) caster sugar

750 ml (1^1/4 pints) crushed ice

Combine the juices. (They should make approximately 250 ml [8 fl oz] of liquid.)

Place the juices and the other ingredients in a food processor or blender and mix until smooth.

The frappé will keep in the freezer for an hour or so.

Serves 2 to 4

Preparation time: 5 minutes

HOT SPIKED LEMONADE

zest of $^1/_2$ lemon, removed in strips

175 ml (6 fl oz) freshly squeezed lemon juice

150 g (5 oz) sugar

7.5 cm (3 in) cinnamon stick

5 tablespoons brandy, Cognac or light rum

very thin slices of lemon to garnish (optional)

Place 600 ml (1 pint) of water and all of the ingredients, except the liquor and lemon slices, in a nonreactive saucepan. Bring the mixture to the boil, stirring to dissolve the sugar. Lower the heat and simmer for 5 minutes.

Remove the pan from the heat. Take the cinnamon stick and the lemon strips out of the pan and discard. Add the liquor and serve garnished with lemon slices, if desired.

Serves 4 to 6

Preparation time: 20 minutes

LEMON AND THYME TEA

2 tablespoons fresh thyme leaves or 1 tablespoon dried thyme

zest of 1 lemon, removed in strips

lemon wedges (optional)

honey (optional)

Place the thyme leaves and lemon zest in a teapot. Add 750 ml (1$^1/_4$ pints) of boiling water and allow to steep for 5 minutes. Strain and serve with lemon wedges and honey, if desired.

For variety, strain and add honey, stirring to dissolve it. Then cool and refrigerate before serving over ice with lemon wedges and thyme sprigs to garnish.

Serves 3 to 4

Preparation time: 10 minutes

SHERRIED PINK GRAPEFRUIT JUICE

500 ml (16 fl oz) freshly squeezed pink grapefruit juice

250 ml (8 fl oz) good-quality dry sherry

ice cubes

pink grapefruit slices or wedges to garnish

Combine the juice and the sherry. Chill for 1 hour or more. Fill four tall glasses with ice cubes and the juice mixture.

Garnish with grapefruit slices and serve.

Serves 4

Preparation time: 5 minutes

GRAPEFRUIT AND PINEAPPLE RUM PUNCH

600 ml (1 pint) freshly squeezed grapefruit juice

350 ml (12 fl oz) fresh or unsweetened canned pineapple juice

100 g (3^1/$_2$ oz) caster sugar

350 ml (12 fl oz) white rum

1 litre (1^3/$_4$ pints) soda water

pineapple and grapefruit slices (optional)

mint leaves (optional)

In a bowl or jug, combine the juices. Add the sugar, stirring to dissolve. Add the rum and stir until well blended. Chill thoroughly.

When ready to serve, gently stir in the soda or water and, if liked, garnish with the fruit slices and mint leaves. Serve 'neat', or over ice.

Serves 6 to 8

Preparation time: 10 minutes

BLOOD ORANGE LIQUEUR

zest of 1 blood orange 6 or 7 blood oranges

500 ml (16 fl oz) vodka

300 g (10 oz) caster sugar

Makes 3 x 500 ml (16 fl oz) bottles.

Preparation time: 30 minutes, plus several months (at least 6) for ageing.

Remove the zest with a vegetable peeler in long strips. Reserve both the zest and the orange. Peel the rest of the oranges and scrape off as much of the white pith from the fruit segments as possible.

With a vegetable knife remove the white pith from the reserved orange, cutting just through the membrane and into the fruit. Spiral down and around the fruit as if peeling an apple. Insert the knife next to the dividing membrane and pop out each segment. Place the orange segments in a food processor or blender and purée briefly. (There should be about 1.25 litres [2 pints] of pulp.) Place the pulp, zest and vodka in a large glass jar. Cover it with greaseproof paper and seal it with a lid, making the jar airtight. Leave the mixture for 1 month in a cool place, shaking gently a few times a week.

When the month is up, strain the mixture through several layers of muslin spread in a sieve. When only pulp remains, pull up the ends of the muslin, enclosing the pulp, and twist it to extract as much liquid as possible. Return the strained liquid to the glass jar. (Make sure the jar has been well rinsed.)

Combine the sugar with 175 ml (6 fl oz) of water in a small pan. Bring to the boil, stirring constantly, being careful not to let it boil over. When all the sugar has dissolved and the syrup is clear, remove from the heat and cool to room temperature. Add the cooled syrup to the vodka and juice mixture. Stir well. Seal again with greaseproof paper and lid. Leave for 2 or 3 days, until the liqueur has cleared and sediment has settled. Siphon or carefully pour the clear portion of the liqueur into smaller bottles. Seal and leave for several months so the liqueur can mellow and age. (Use the remaining portion with the sediment for cooking, or drink it yourself.)

RECIPE LIST

Baked Fish with Ugli Fruit
60

Beef with Tangerines
74

Blood Orange Liqueur
122

Buckwheat Pancakes with Clementines
32

Chicken Salad with Tangerines
and Red Onions
48

Chicken with Lemon and Olives
64

Chocolate-Dipped Citrus Segments
92

Chocolate and Orange Marmalade Brownies
98

Citrus Frappé
116

Citrus Marmalade
80

Curried Orange Soup
42

Pried Scallops with Lime and Garlic
56

Frozen Blood Orange Soufflé
104

Grapefruit-Marinated Poussins
62

Grapefruit and Pineapple Rum Punch
120

Grilled Pink Grapefruit and Pork Skewers
36

Grilled Shark Steaks with Citrus Salsa
58

Hot Spiked Lemonade
118

Iced Orange Coffee
114

Kumquat and Ginger Preserve
82

Lemon and Allspice Muffins
24

Lemon Fettuccine with Peppered Prawns
54

Lemon and Nutmeg Cheesecake
106

Lemon and Parsley Soup
40

Lemon Snaps
96

Lemon Steam Cake
with Blood Orange Sauce
108

Lemon and Thyme Tea
118

Limeade Fizz
116

Lime and Tomato Relish
84

Lime Mousse Tart
102

Lime Soup
38

Mandarin Orange Upside-Down Cake
110

Maple and Orange Ambrosia
90

Minted Ruby Grapefruit Ice
94

Navel Orange Waffles with Blueberry Sauce
28

Orange and Banana Muffins
24

Orange and Coconut Custard
100

Orange Curd
86

Orange Pain Perdu
30

Pickled Citrus Prawns
50

Pickled Lemons
88

Pink Grapefruit Poached in Sauternes
20

Pink Lemonade
114

Pork Medallions with Mandarin Oranges
and Cranberries
70

Salad of Beetroot, Lamb's Lettuce and
Clementines
44

Sherried Pink Grapefruit Juice
120

Soba with Fennel, Chèvre and Kumquats
66

Spareribs with Lime and Chilli
68

Stewed Kumquats and Strawberries
22

Tangelo and Onion Relish
84

Tangerine and Cranberry Chutney
86

Tangerine-Pecan Scones
26

Tangerine Sorbet
94

Ugli Fruit Caribbean Fish Salad
46

Veal Chops with Blood Oranges
72

RECIPE LIST BY CITRUS

CLEMENTINES

Buckwheat Pancakes with Clementines
32

Salad of Beetroot, Lamb's Lettuce
and Clementines
44

CITRUS

Chocolate-Dipped Citrus Segments
92

Citrus Frappé
116

Citrus Marmalade
80

Grilled Shark Steaks with Citrus Salsa
58

Pickled Citrus Prawns
50

GRAPEFRUIT

Grapefruit-Marinated Poussins
62

Grapefruit and Pineapple Rum Punch
120

Grilled Pink Grapefruit and Pork Skewers
36

Minted Ruby Grapefruit Ice
114

Pink Grapefruit Poached in Sauternes
20

Sherried Pink Grapefruit Juice
120

KUMQUATS

Kumquat and Ginger Preserve
82

Soba with Fennel, Chèvre and Kumquats
66

Stewed Kumquats and Strawberries
22

LEMONS

Chicken with Lemon and Olives
62

Hot Spiked Lemonade
118

Lemon and Allspice Muffins
24

Lemon Fettuccine with Peppered Prawns
54

Lemon and Nutmeg Cheesecake
106

Lemon and Parsley Soup
40

Lemon Snaps
96

Lemon Steam Cake with
Blood Orange Sauce
108

Lemon and Thyme Tea
118

Pickled Lemons
78

Pink Lemonade
116

LIMES

Fried Scallops with Lime and Garlic
56

Limeade Fizz
116

Lime and Tomato Relish
84

Lime Mousse Tart
102

Lime Soup
38

Spareribs with Lime and Chilli
68

ORANGES

Chocolate and Orange Marmalade Brownies
98

Curried Orange Soup
42

Frozen Blood Orange Soufflé
104

Iced Orange Coffee
114

Mandarin Orange Upside-Down Cake
110

Maple and Orange Ambrosia
90

Navel Orange Waffles with Blueberry Sauce
28

Orange and Banana Muffins
24

Orange and Coconut Custard
100

Orange Curd
86

Orange Pain Perdu
30

Pork Medallions with Mandarin Oranges
and Cranberries
70

Veal Chops with Blood Oranges
72

TANGELOS

Tangelo and Onion Relish
84

TANGERINES

Beef with Tangerines
74

Chicken Salad with Tangerines and Red
Onions
48

Tangerine and Cranberry Chutney
86

Tangerine and Pecan Scones
26

Tangerine Sorbet
94

UGLI FRUIT

Baked Fish with Ugli Fruit
60

Ugli Fruit Caribbean Fish Salad
46

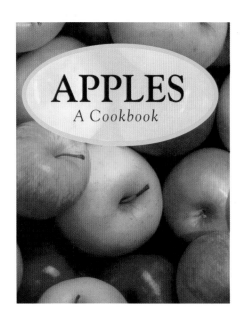

APPLES
A Cookbook

BERRIES
A Cookbook

NUTS
A Cookbook

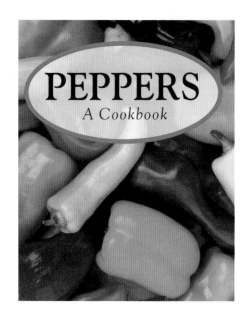

PEPPERS
A Cookbook